RICE

Lynne Merrison

Illustrations by John Yates

Carolrhoda Books, Inc./Minneapolis

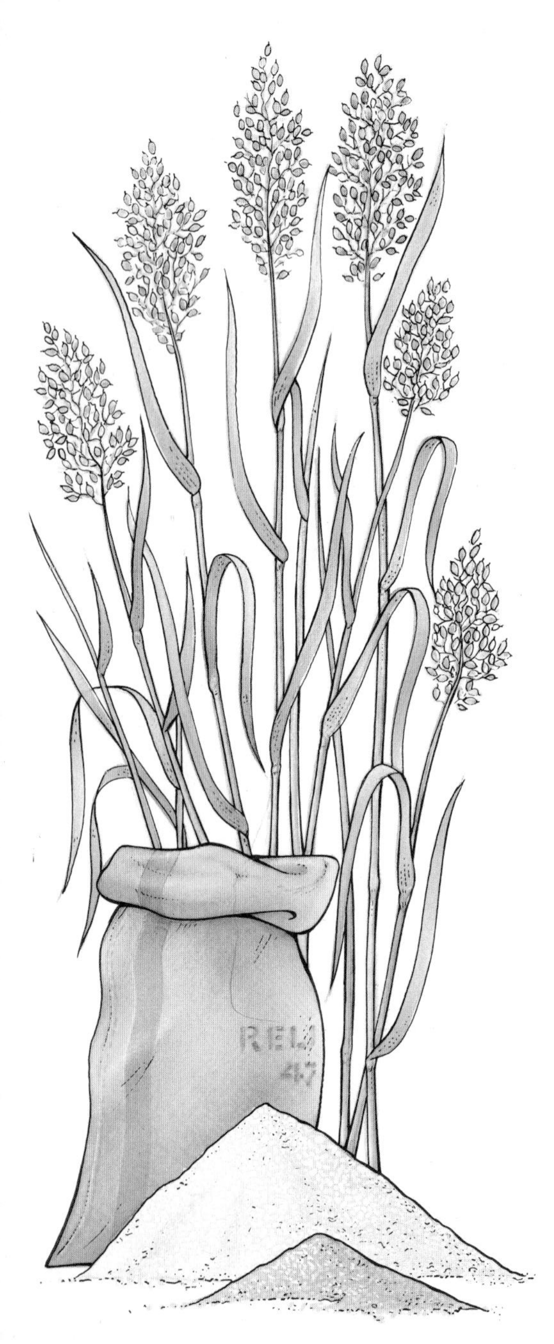

All words that appear in **bold** are
explained in the glossary on page 30.

First published in the U.S. in 1990 by
Carolrhoda Books, Inc.
 First published 1989 by Wayland (Publishers) Ltd.

Library of Congress Cataloging-in-Publication Data

Merrison, Lynne.
Rice / Lynne Merrison ; illustrations by John Yates.
p. cm. — (Food we eat)
Summary: Describes rice, the history of its cultivation, and the part it plays in human health and diet. Includes several recipes.
ISBN 0-87614-417-2 (lib. bdg.)
1. Rice—Juvenile literature. 2. Cookery (Rice)—Juvenile literature. [1. Rice. 2. Cookery—Rice.] I. Yates, John, ill. II. Title. III. Series: Foods we eat (Minneapolis, Minn.)
SB191.R5M446 1990
641.3′318—dc20 89-25119
CIP
AC

Printed in Italy by G. Canale C.S.p.A., Turin
Bound in the United States of America

1 2 3 4 5 6 7 8 9 10 99 98 97 96 95 94 93 92 91 90

Contents

What is rice?

There are over seven thousand known varieties of rice. Most of us have eaten either long-grain, short-grain, or brown rice. Like wheat, oats, and corn, rice is a **cereal**. It provides a basic diet for more than half of all the people in the world. In fact, rice is considered to be one of the most important crops grown by the world's farmers. It is especially important in some developing countries, such as India, where the warm, wet climates are ideal for growing rice.

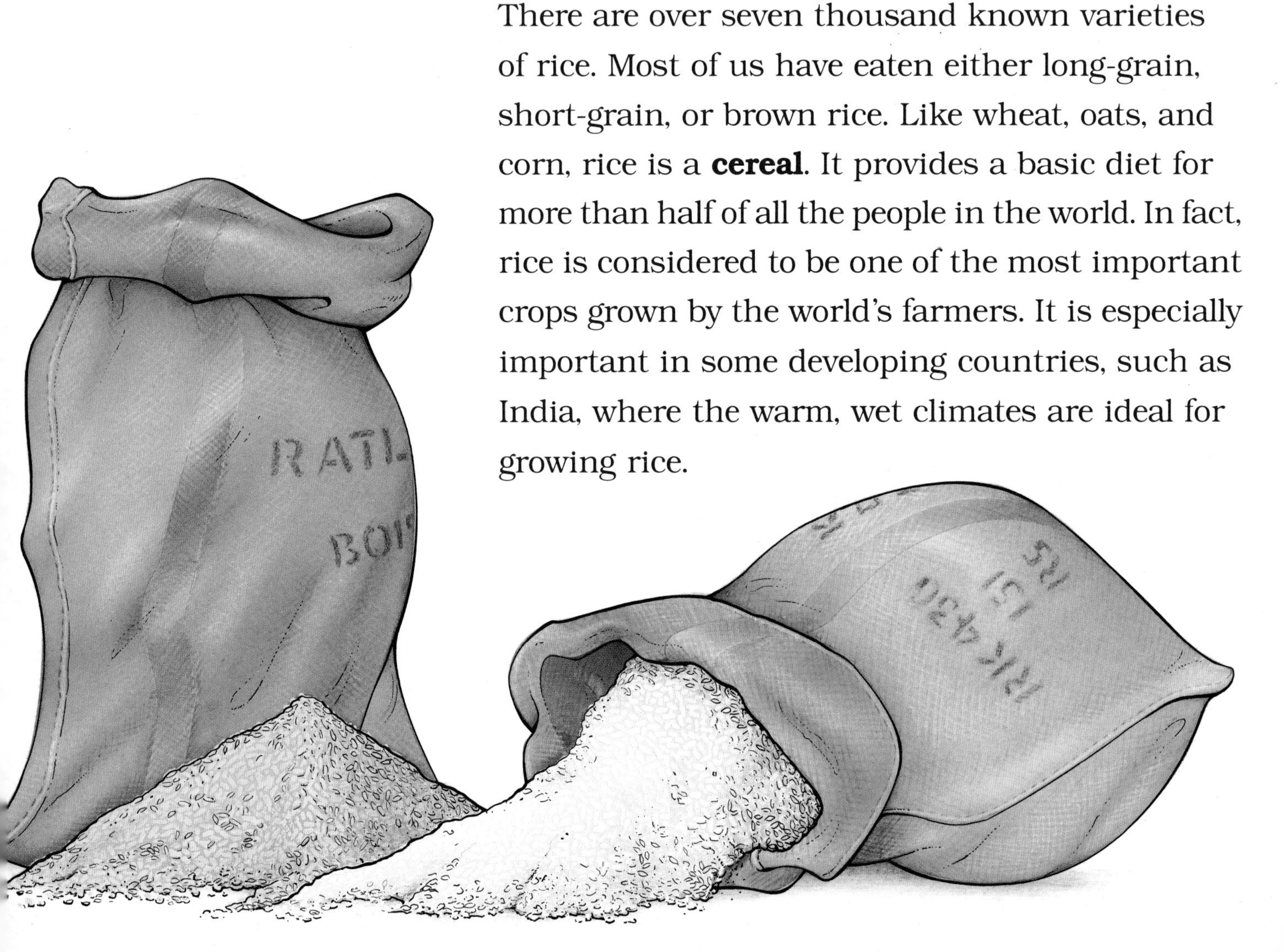

Rice is rich in **carbohydrates**, which are an important part of the human diet. But to stay healthy, we also need **protein**. Rice does contain some protein, but not enough for a healthy, balanced diet. In some parts of the world, however, people eat only rice. These people sometimes suffer from malnutrition, caused by a lack of essential **nutrients**.

These are some of the many different types of rice eaten by people all over the world.

Rice in the past

It is thought that wild rice seeds were first collected and planted by prehistoric people in Southeast Asia over seven thousand years ago. It was grown for food in the watery deltas of the areas that are now Bangladesh, Burma, and Vietnam. From these places, traders carried seeds to the areas that are now China and Indonesia. Rice may have been grown in China as early as 5000 B.C.

Invading armies and traveling merchants

Preparing a paddy field for planting in Japan in 1890

introduced rice to the rest of Asia and gradually to the Middle East and southern Europe. In the 17th century, rice was brought to the United States. The United States eventually became the world's largest exporter of rice.

This illustration shows a rice harvest in Hong Kong in the 19th century.

Rice as food

Rice is easy to digest and can be used in a variety of ways, making it one of the most popular and widely eaten foods in the world. It can be eaten on its own, as part of a salad, or along with another dish. There are many different types and flavors of rice.

Rice is a nutritious part of our diets, especially

Chinese children enjoying a delicious meal of rice

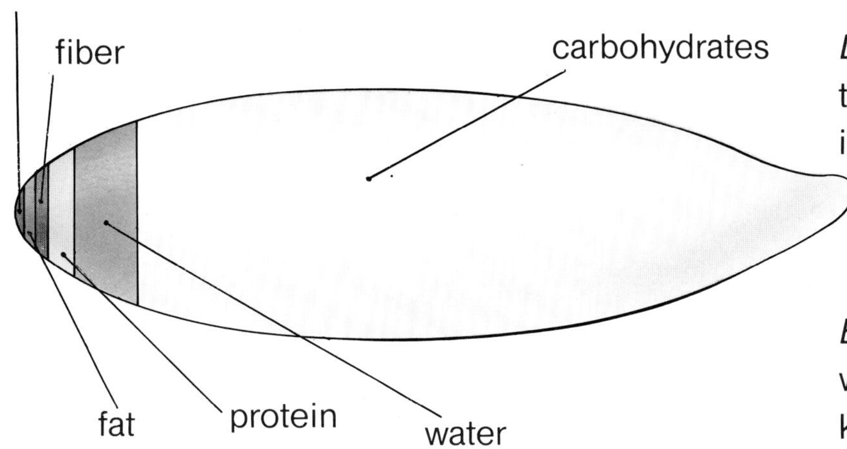

Left: The amounts of the substances found in a grain of rice

Below: Rice goes well with dishes such as kebabs and curry.

when eaten with foods that contain protein, such as fish, meat, and beans. Much of the food value of rice, however, is lost in the **milling** process that produces the white rice commonly found in stores. The first step in milling rice is **winnowing**, which removes the hard outer **hull** from the rice grain. **Polishing** then removes the soft layers of **bran** that are left, as well as some of the **endosperm** underneath. Many people prefer white rice because it takes less time to cook than unpolished brown rice. Brown rice, however, is more nutritious than white rice.

Where does rice come from?

Rice comes from rice plants that are usually grown in flooded fields called **paddies**. In paddies, the roots of the rice plants are kept under water. The head of a rice plant is called a **panicle**. It grows on a stalk above long, flat leaves. The panicle has small flowers called **spikelets** that produce the grains of raw rice.

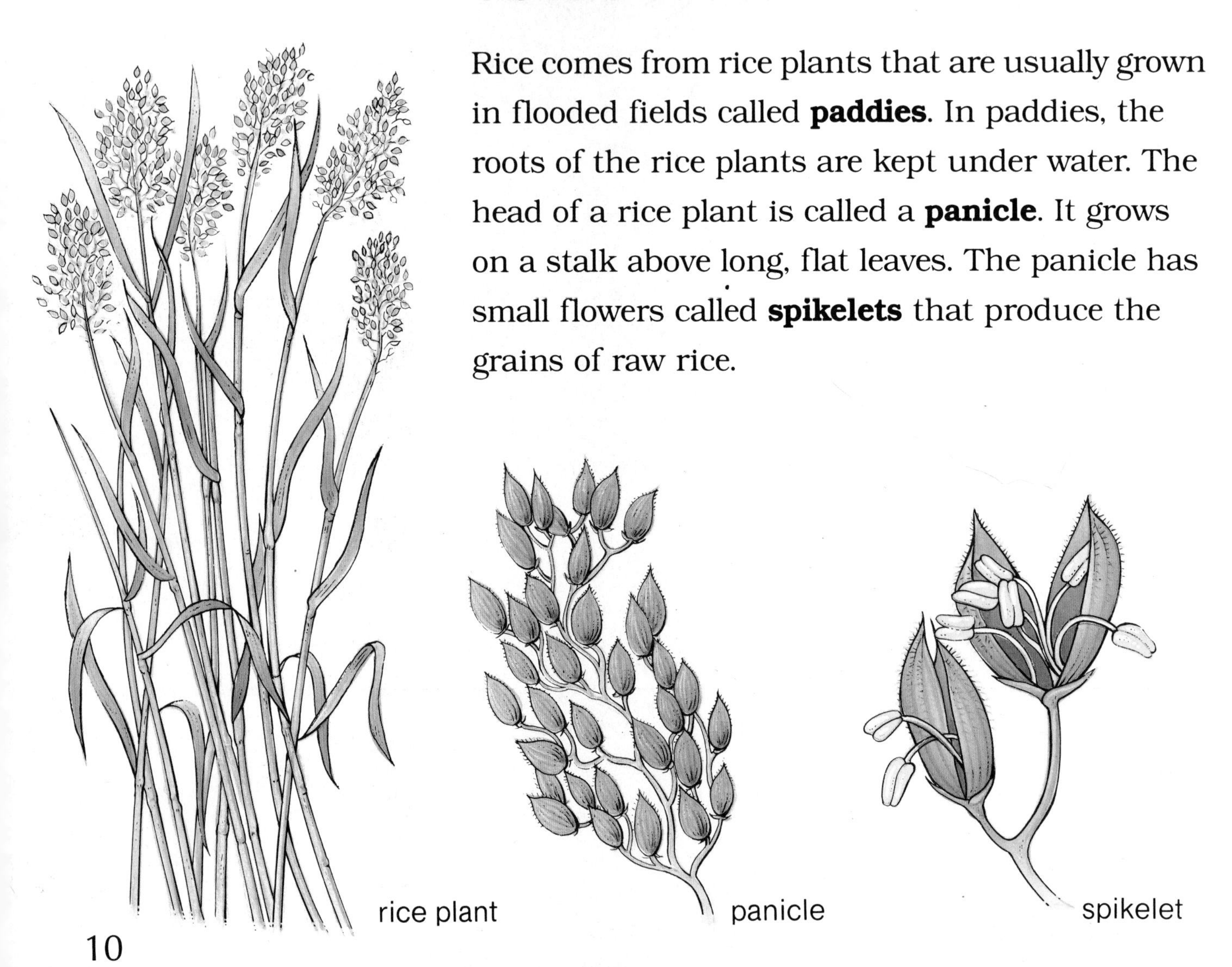

Right: Rice fields, or paddies, that are laid out in terraces in the Philippines

Below: The layers of a grain of rice

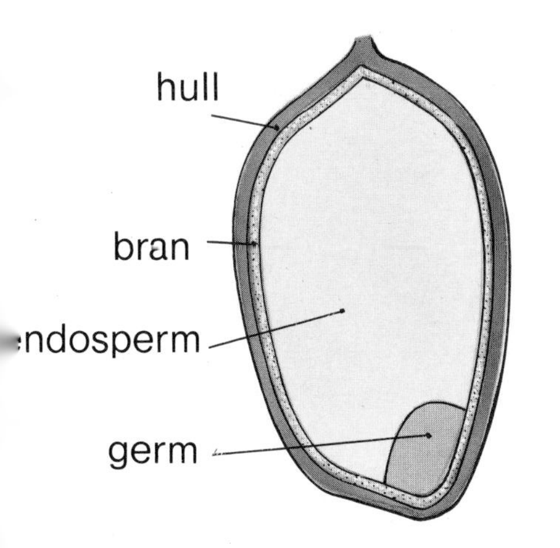

Rice is grown in about one hundred countries. It needs plenty of water and sunshine, so tropical and subtropical climates are ideal for growing rice. The monsoon regions of Southeast Asia are especially well suited to rice farming. China, India, and other Asian countries produce about 90 percent of the world's rice. It is also grown in parts of the United States, South America, Europe, and Africa.

Growing rice

Rice can grow in many types of climates and terrain, but most of the world's rice is grown in paddies. Before fields are flooded to make the paddies, the soil has to be broken up and leveled. Some farmers use cattle such as buffalo to help plow the paddy

A rice farmer in Indonesia using a buffalo to help plow his paddy field

This Japanese rice farmer has planted rice seeds in a nursery bed. In about 30 days she will plant them in the paddies.

fields. In some countries, the fields are plowed by machines.

In some countries, rice seeds are planted in the paddies by machines. In most of the countries where rice is grown, however, the seeds are first sown by hand in **nursery beds**. These are small sections of the paddies that are specially set aside for sowing new plants. For their first night, the rice seeds are kept in water. This allows them to

germinate, or grow small shoots. The shoots are then planted in the nursery beds.

It usually takes about 30 days before the young shoots are ready to be planted in the paddies. By

Above: Planting rice shoots in mud

Right: Preparing a paddy field for planting

this time, the fields may need to be plowed again to remove any weeds that have grown. Planting rice is hard, tiring work. The rice plants have to be pressed into the mud in neat rows. For the people planting the seeds, this means standing in water and bending down for long periods of time.

Rice farmers work hard to keep their crops healthy. As the young shoots begin to grow, they are endangered by pests, such as rats, worms, and birds, and by weeds. Snails that live in the muddy water can also harm a rice crop. Many farmers keep fish in their paddy fields. The fish help protect the rice crop by eating worms and insects. They also provide the farmers with another source of food.

Spraying **pesticides** on the crop can also help protect it from insects and diseases. These chemicals are usually expensive, however, and many rice farmers cannot afford them.

Hurricanes, drought, and other natural forces can also damage or destroy rice crops.

Removing weeds from a rice paddy in Canton, China

Harvesting rice

When rice is ready for harvesting, it is a golden yellow color. In some countries where rice is grown, rice plants are harvested by a machine called a combine harvester. Most harvesting, however, is done by hand, using a sharp knife or a sickle. After the rice plant stalks are cut, farmers arrange them in bundles, or **sheaves**. The sheaves are tied up and left out in the fields, where they are dried by the sun.

Above: Threshing rice using a foot-operated machine in Katmandu, Nepal

Right: Threshing rice with a special type of machine

Rice can be winnowed by being tossed in the air from a sieve. The wind carries away the hulls.

After harvesting, the grain has to be separated from the straw. This process is called **threshing**. Combine harvesters can cut the rice, separate the grain from the straw, and leave the straw in the field quickly and easily. In most countries where rice is grown, however, threshing is done by hand.

The rice that is left after threshing still has the hard outer layer called the hull and the inner layers of bran. Rice that still has the hull and bran is called **rough rice** or **paddy rice.** This rice must be milled before it can be packaged and sold.

This machine separates the hull and bran from the rice grains. The rice is then polished to remove any remaining layers of bran.

Milling rice

After being threshed and dried, rice is winnowed to remove the tough outer hull. Some rice then goes through the polishing process, which removes the bran from the edible rice grains. Unfortunately, the bran and the hull contain most of the nutrition in the rice. White rice is sometimes parboiled—soaked in water and then cooked briefly—before the bran is removed. This puts some of the nutrition from the bran into the rice grains.

Most countries have their own methods of winnowing and polishing rice by hand. One common method of winnowing rice by hand is to pound the grain in a wooden bowl, using long poles. It is then cleaned again by being rubbed through a sieve. This removes any remaining bits of hull. In some countries, rice is milled by machine in commercial rice mills.

Rice can be milled by hand. The grain is pounded with long poles in wooden bowls.

Cooking and eating rice

Rice is eaten in almost every country of the world. Spanish paella contains rice, as does jambalaya, a spicy dish eaten in the southern United States. Rice can be eaten as a side dish with fish, meat, and poultry. It is the main ingredient in dishes such as risotto, which is eaten in Italy, and pilaf, a Middle Eastern dish. Rice can be mixed with fruit or vegetables to make salads.

A delicious chicken and rice dish

Left: These are pieces of sushi, a traditional Japanese dish of vegetables or raw fish, seaweed, and rice.

Below: Paella—chicken, seafood, and rice—is a popular Spanish dish.

There are many other ways of eating rice. It can be toasted or coated with sugar or chocolate to make breakfast cereal. It is used to make rice pudding, a popular dessert. Rice is also used in soups and in many baby foods. Seasoned rice cakes and crackers are eaten as snacks in many countries.

Rice can be eaten on its own, colored with spices such as saffron, or flavored with different seasonings. Scented or aromatic rice is eaten in some countries. Pilau rice, flavored with peppers and onions, is popular in India. Basmati rice is a richly flavored long-grain rice grown in Pakistan and India. Edible rice "paper," made from rice flour, is used in cakes and to wrap candies. Rice can also be used to produce many different drinks. The best

Above: Rice balls are a typical Asian dessert.

Right: This young boy is enjoying a bowl of rice.

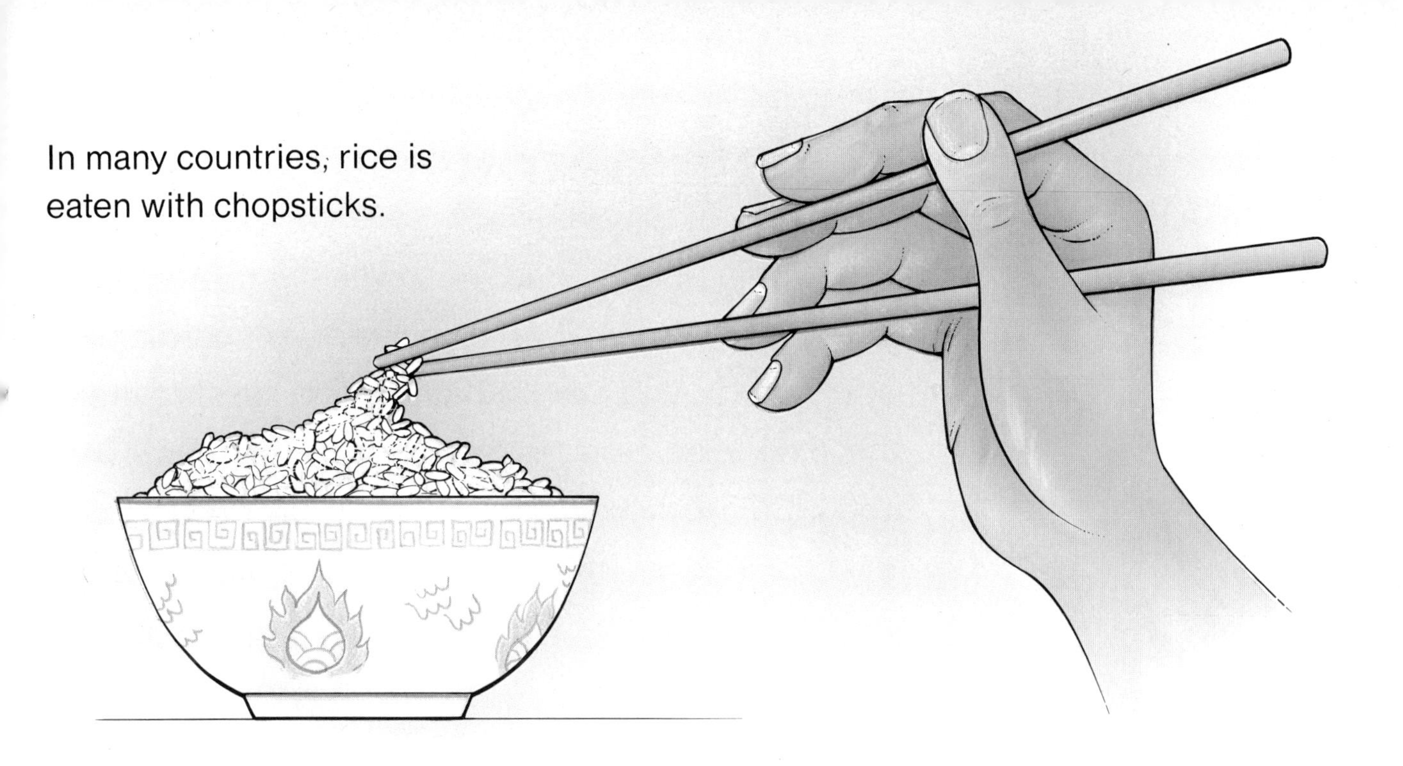
In many countries, rice is eaten with chopsticks.

known alcoholic drink made from rice is **sake**, which is made in Japan. About one-third of the rice used in the United States is found in beer.

Most Americans cook rice by boiling it in water. It absorbs the water and expands to almost three times its original size. Rice can also be steamed, which is the healthiest method of cooking rice. Rice that has been boiled or steamed and then fried is popular in China.

Below: Rice is used in many breakfast cereals.

Beliefs about rice

Rice has played such an important role in the lives of people in Asia that it has become part of many of the cultures and religions there. It is the main crop in Asia and a vital part of the diet of many Asian people. In Indonesia, some people worship the Rice Mother. They believe that the first grains of rice were produced in the Rice Mother's body, and that she watches over all the crops. Festivals celebrating the budding and harvesting of rice

Preparing gifts for the rice gods in Indonesia

Left: Rice and other foods are offered to the rain gods in the hope that they will help the crops grow.

The *pongal* ceremony in India is to celebrate and give thanks for a good rice harvest.

plants are still held in parts of Indonesia. In Japan, rice is such an important food that the word for rice also means "meal." Members of the Shinto religion in Japan place rice cakes at shrines as offerings to the rice god. Rice is also offered to the gods in parts of India.

Rice plays a part in Western culture as well. Many Europeans and Americans throw rice at the end of a wedding ceremony, a tradition that may have begun in India. Rice festivals are also held in some U.S. cities at harvest time.

Make sure there is an adult nearby when you are cooking.

Pilaf

You will need:

1 bunch of green onions
1 small yellow onion
1 red pepper
1 green pepper
1 carrot
½ cup mushrooms
1 tablespoon cooking oil
a pinch of salt
a pinch of pepper
¾ cup brown rice

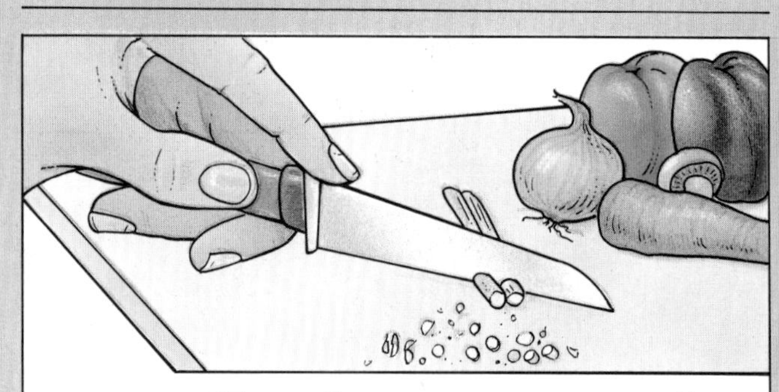

1. Chop the green onions, yellow onion, red and green peppers, carrot, and mushrooms into small pieces.

2. Fry the chopped vegetables in the oil for five minutes. Add the salt and the pepper.

3. Add the rice and fry for one minute. Add 2 cups of water, cover the pan, and simmer for 30 minutes until the water is absorbed.

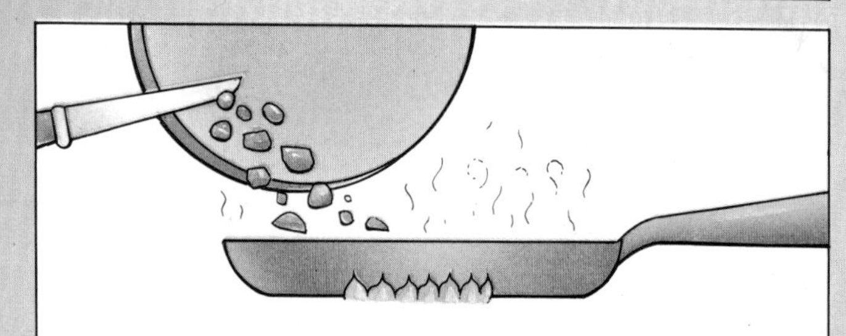

4. You can add pieces of cooked meat or chicken if you like.

Spanish rice

You will need:

½ cup long-grain rice
4 slices of bacon
1 onion, chopped
½ green pepper, chopped
1 12-ounce can tomato soup
4 tablespoons water
salt
pepper

1. Cook the rice in boiling water for 15 minutes and rinse it under cold water. Ask an adult to help you with this part.

2. Chop the bacon and fry until crisp. Remove the bacon from the pan and fry the onion and pepper in the bacon fat until tender.

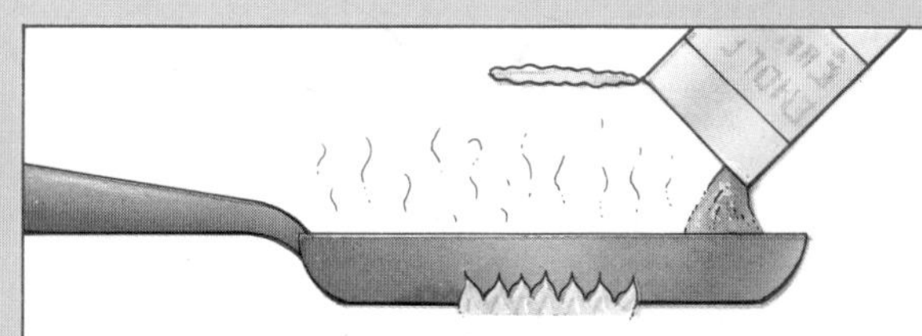

3. Add the bacon, rice, soup, water, salt, and pepper to the pan and cook until heated through.

4. Serve the rice with crackers if you like.

Rice salad with pork

You will need:

½ cup brown rice
10 ounces cooked pork, chopped
¼ cup rolled oats
4 tablespoons milk
2 tablespoons tomato sauce
1 teaspoon mustard
1 teaspoon butter
1 head of lettuce

1. Cook the rice in boiling water for 30 minutes. Rinse under cold water. Ask an adult to help you with this part.

2. Mix together the chopped pork, oats, milk, tomato sauce, and mustard.

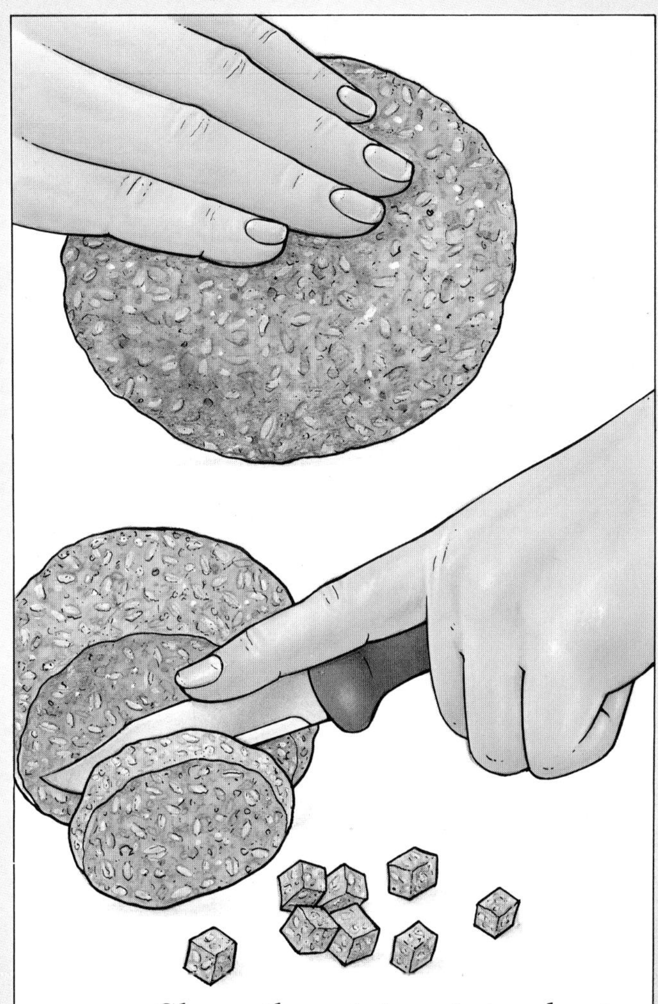

3. Shape the mixture into a large ball and then cut it up into small cubes. Fry the cubes in the butter and then mix with the rice.

4. Serve the rice mixture on a bed of lettuce leaves.

Glossary

bran: the soft coating between the outer hull and the inner seed of a plant

carbohydrates: substances in some foods, such as rice, that are an important source of energy

cereal: a type of plant that produces edible grains

endosperm: the inner part of the seed of a plant

germinate: to produce plant shoots

hull: the hard outer covering of the seed of a plant

milling: the process of making rice grains into the different types of rice that people eat

nursery beds: areas of rice paddies that are specially set aside for sowing new rice plants

nutrients: various substances in food that keep us healthy

paddies: flooded fields in which rice is grown

paddy rice: rice that has been harvested and threshed but still has its hull and bran. Also called rough rice.

panicle: the flowering head of a rice plant

pesticides: chemicals that are sprayed on plants to kill insects and bacteria

polishing: the process of removing the bran from rice grains

protein: a substance found in some foods that is an essential part of human diets

rough rice: rice that has been harvested and threshed but still has its hull and bran. Also called paddy rice.

sheaves: bundles of harvested plants

spikelets: the flowers of the rice plant, which produce rice grains

threshing: separating rice grains from the rest of the rice plant

winnowing: the process of removing the hull from rice grains

Index

Photo Acknowledgments

The photographs in this book were provided by: p. 5, Topham Picture Library, pp. 6, 7, Mary Evans Picture Library; pp. 8, 9, 11, 13, 14 (left), 18, 22 (right), ZEFA; pp. 12, 16 (right), 17, 25 (right), Hutchison Library; p. 14 (right, Brian J. Coates), 15 (Norman Myers), 19 (Michael Freeman), 21 (left, Jonathan T. Wright), 24 (Norman Myers), 25 (left, Norman Myers), Bruce Coleman; p. 16 (left), David Cumming; p. 21 (right), Wayland Picture Library; p. 22 (left), Christine Osborne.